INTEGRATED SCIENCE LAB MANUAL

FOR CSEC

Shonique Ebanks

(BSc., MSc., Dip. Ed.)

CSEC® is a registered trade mark of the Caribbean Examination Council (CXC). Integrated Science Lab Manual For CSEC is an independent publication and is not authorised, sponsored, or otherwise approved by CXC.

Table of Contents

Experiment # 1:   Diffusion

Experiment # 2:   Osmosis

Experiment # 3:   Photosynthesis 1

Experiment # 4:   Photosynthesis 2

Experiment # 5:   Food and Nutrition

Experiment # 6:   Enzyme Activity 1

Experiment # 7:   Enzyme Activity 2

Experiment # 8:   Respiration 1

Experiment # 9:   Respiration 2

Experiment # 10:   Transpiration

Experiment # 11:   Circulation 1

Experiment # 12:   Circulation 2

Experiment # 13:   Excretion

Experiment # 14:   Coordination

Experiment # 15:   Sexual Reproduction 1

Experiment # 16:   Sexual Reproduction 2

Experiment # 17:   Asexual Reproduction 1

Experiment # 18:   Asexual Reproduction 2

Experiment # 19:   Asexual Reproduction 3

Experiment # 20:   Conduction

Experiment # 21:   Radiation

Experiment # 22:   Terrestrial Environment 1

Experiment # 23:   Terrestrial Environment 2

Experiment # 24:	Aquatic Environment 1
Experiment # 25:	Aquatic Environment 2
Experiment # 26:	Sanitation
Experiment # 27:	Safety Hazards
Experiment # 28:	Chemical Reactions of Metals
Experiment # 29:	Corrosion
Experiment # 30:	Acid, Bases and Salt
Experiment # 31:	Hardness of Water
Experiment # 32:	Soap and Soapless Detergents
Experiment # 33:	Current, Voltage and Resistance

Experiment #: 1

Title: Diffusion

Aim: To investigate diffusion of potassium permanganate (VII) in water.

Skill: Observation, Recording and Reporting

Materials: beaker, water, potassium permanganate (VII)

Procedure:

1. Carefully drop some potassium permanganate (VII) crystals into the beaker by letting it slide down the side of the glass.
2. Observe what happens.
3. Leave the beaker undisturbed for an hour and describe what has happened.
4. Leave the beaker undisturbed overnight and then describe its appearance.

Questions:

1. What is diffusion?
2. Explain the observations you made in the table.
3. How could the rate of diffusion in the experiment above be increased?

Conclusion

Name: _____   Date: _____

Form: _____   Lab Partner: _____

Experiment #: 1

Title: Diffusion

Results:

Title: _____

Test of potassium permanganate (VII)	Observation
Just added	
After 1 hour	
After overnight	

Teacher's Signature: _____   Date: _____

Experiment #: 2

Title: Osmosis

Aim: To investigate osmosis in potato strips

Skills: Observation, Recording and Reporting; Manipulation and Measurement; Analysis and Interpretation

Materials: potato, scalpel, distilled water, beakers, salt water, ruler, stop watch

Procedure:

1. Use the scalpel to peel the potato.
2. Cut 2 strips of potato 4 cm in length.
3. Place one of the potato strip in distilled water and the other strip in salt water.
4. Measure the length of both potato strips at every 3 minutes interval for a total of 15 minutes.
5. Note the texture of the potato strips at the end of the experiment.
6. Record the results in a table.
7. Plot a graph of Length of Potato (cm) against Time (min).

Questions:

1. Define osmosis.
2. Why was there a difference in texture in the potato strips?
3. Use the graph and your knowledge of osmosis to explain the change in length of the potato strips.

Conclusion

Name: _____   Date: _____

Form: _____   Lab Partner: _____

Experiment #: 2

Title: Osmosis

Results:

Title: _____

Time / min	Length of potato in fresh water (cm)	Texture	Length potato in salt water (cm)	Texture
3				
6				
9				
12				
15				

Teacher's Signature: _____   Date: _____

Experiment #: 3

Title: Photosynthesis 1

Aim: To show the external structure of a hibiscus leaf

Skill: Drawing

Materials: hibiscus leaf, white paper, pencil, ruler

Procedure:

1. Pick a hibiscus leaf.
2. Place the leaf flat on the table with the smooth side facing upwards.
3. Border the white paper by drawing a 1 cm line from the edge of the paper.
4. Draw the outline of the leaf and include the major veins.
5. Label the main parts of the leaf to the right side of the drawing.
6. Give your drawing a title.
7. Measure the length of the leaf (on the table).
8. Measure the length of your drawing.
9. Calculate the magnification of the drawing.

Name: _____   Date: _____

Form: _____

Experiment #: 3

Title: Photosynthesis

Teacher's Signature: _____   Date: _____

Experiment #: 4

Title: Photosynthesis 2

Aim: 1. To test the presence of starch in leaf.

2. To test if chlorophyll is necessary for photosynthesis.

Skills: Observation, Recording and Reporting; Analysis and Interpretation

Materials: beaker, water, Bunsen burner, tripod stand, wire gauze, boiling tube, iodine, dropper, ethanol, variegated leaf, green leaf, evaporating dish, white tile

Procedure:

1. Dip both leaves in boiling water for about 1 minute.
2. Turn off the burner so there is no flame.
3. Put the both leaves into a separate boiling tube containing ethanol to cover it. Place the boiling tube in the beaker of hot water and leave it for 10 minutes.
4. Take the leaf out of the ethanol and wash it in cold water.
5. Note the appearance of both leaves.
6. Spread the leaves out in an evaporating dish or a white tile and cover it with iodine solution.
7. Note the colour change of both leaves.

Discussion: Your discussion should have the following explained:

- ✓ First paragraph: define photosynthesis; equation for photosynthesis; the conditions needed for photosynthesis.
- ✓ Second paragraph: explain why the leaf was boiled for 1 minute; give the purpose of using the ethanol on the leaf; give the purpose of using iodine on the leaf; explain the colour of the leaves after iodine was added; why did some parts of the variegated leaf changed colour and some parts didn't?

Conclusion

Name: _____   Date: _____

Form: _____   Lab Partner: _____

Experiment #: 4

Title: Photosynthesis

Results:

Title: _____

Appearance	Green Leaf	Variegated Leaf
After washing with ethanol		
After iodine solution		

Teacher's Signature: _____   Date: _____

Experiment #: 5

<u>Title:</u> Food and Nutrition

<u>Aim:</u> To test the presence of starch, reducing sugars, non-reducing sugars, protein and lipids in food samples.

<u>Skills:</u> Observation, Recording and Reporting; Manipulation and Measurement; Analysis and Interpretation

<u>Test for starch</u>

Aim: To test for the presence of starch

Materials: iodine solution, dropping pipette, test tubes

Procedure: 1. Pour 2ml of the food in a test tube.

2. Using a dropping pipette, add 3 drops of iodine solution.

* Positive colour change – yellow- orange (iodine) to blue-black

<u>Test for reducing sugars</u>

Aim: To test for the presence of reducing sugars

Materials: Benedict's solution, copper sulphate (alkaline), Bunsen burner, test tubes

Procedure: 1. Pour 2 ml of the food in a test tube.

2. Add 2 ml of Benedict's solution, followed by an equal volume of copper sulphate solution.

3. Boil in a water bath (do not heat directly with a Bunsen burner).

4. Watch carefully for colour changes.

* Positive colour change – blue (Benedict's solution) to green/yellow/orange/red

Test for non-reducing sugars

Aim: To test for the presence of non-reducing sugars

Materials: Benedict's solution, dilute hydrochloric acid, Bunsen burner, test tubes, dilute sodium hydroxide

Procedure: 1. Pour 1ml of the food in a test tube.

2. Add a few drops of dilute hydrochloric acid to the sample and boil for a few minutes.

3. Cool the test tube and add dilute sodium hydroxide solution. (Beware, the latter will fizz).

4. When neutralised, test with 1ml of Benedict's solution.

* Positive colour change – blue (Benedict's solution) to green/yellow/orange/red

Test for proteins

Aim: To determine the presence of protein

Materials: biuret solution (a solution of copper sulphate and sodium hydroxide), test tubes

Procedure: 1. Pour 1ml of the food in a test tube.

2. Add 1 ml of biuret solution and mix by shaking the tube from side to side.

* Positive colour change – blue (biuret solution) to violet/purple/lilac

Test for lipids

Aim: To determine the presence of lipid

Materials: ethanol, water, test tubes

Procedure:
1. Pour 1ml of the food in a test tube.
2. Add 1 ml of ethanol and shake to dissolve.
3. Pour off the ethanol, which may have dissolved some lipids, into a test tube of water (do not mix).

* Positive colour change – white cloudiness – an emulsion

Discussion:

- ✓ Discuss the importance of starch, reducing sugars, non-reducing sugars, protein and lipids to the human diet and body (in a separate paragraph). Also, in each paragraph give the food samples that were positive for the different tests.

Conclusion

Name: _____   Date: _____

Form: _____   Lab Partner: _____

Experiment #: 5

Title: Food and Nutrition

Results:

Title: _____

Test	Food Samples				
Starch					
Reducing sugars					
Non-reducing sugars					
Protein					
Lipid					

Key:

✓	Present
x	Absent

Teacher's Signature: _____   Date: _____

Experiment #: 6

Title: Enzyme Activity 1

Aim: To plan and design an experiment to investigate the rate of reaction of salivary amylase over the temperature range 10 – 50 °C.

Skill: Plan and Design

Hypothesis:

Materials:

Variable: Manipulating:

  Controlling:

  Responding:

Procedure:

Expected Results:

Treatment of results:

Source of error:

Experiment #: 7

Title: Enzyme Activity 2

Aim: To investigate the effect of pH on enzyme activity

Skills: Observation, Recording and Reporting; Manipulation and Measurement; Analysis and Interpretation

Materials: starch solution, beaker, test tubes, dilute hydrochloric acid, dilute sodium hydroxide, measuring cylinder, salivary amylase solution, iodine solution, stop watch, spotting tile, dropper

Procedure:

1. Pour some starch solution into a beaker.
2. Measure 3 cm³ of the starch solution and pour into three test tubes, A, B and C.
3. Add three drops of dilute hydrochloric acid to test tube A.
4. Add three drops of dilute sodium hydroxide solution to test tube B.
5. Add 1 cm³ of salivary amylase solution to each of test tubes A, B and C.
6. After 2 minutes place a drop of solution from each test tube on a spotting tile and add one drop of iodine solution. Record your observations.
7. Repeat step 6 at 2 minute intervals for 20 minutes.
8. Record your results in a table.

Discussion:

- ✓ First paragraph: Define enzymes; why are enzymes important;
- ✓ Second paragraph: state how pH affects enzyme activity; explain how the hydrochloric acid and sodium hydroxide affect the enzyme from the results; explain the change in colour when iodine was added over the 20 minutes

Conclusion

Name: _____ Date: _____

Form: _____ Lab Partner: _____

Experiment #: 7

Title: Enzyme Activity

Results:

Title: _____

Time / min	A	B	C
2			
4			
6			
8			
10			
12			
14			
16			
18			
20			

Teacher's Signature: _____ Date: _____

Experiment #: 8

<u>Title:</u> Respiration 1

<u>Aim:</u> To investigate the amount of heat produced by germinating seeds.

<u>Skills:</u> Observation, Recording and Reporting; Analysis and Interpretation

<u>Materials:</u> two vacuum flask, two thermometers, two clamps and stands, cotton wool, germinating peas, germinating peas that have been boiled and cooled, dilute disinfectant solution

<u>Procedure:</u>

1. Rinse both sets of germinating peas in dilute disinfectant solution to kill microbes and fungi.
2. Pour the germinating peas into one of the vacuum flasks, insert the thermometer and plug the top with cotton wool.
3. Secure the flask with the clamp.
4. Repeat the second flask, filling it with the boiled peas and inserting the thermometer and cotton wool plug.
5. Record the thermometer in each flask.
6. Leave the flasks for 24 hours and record the temperature in each flask.

<u>Diagram:</u>

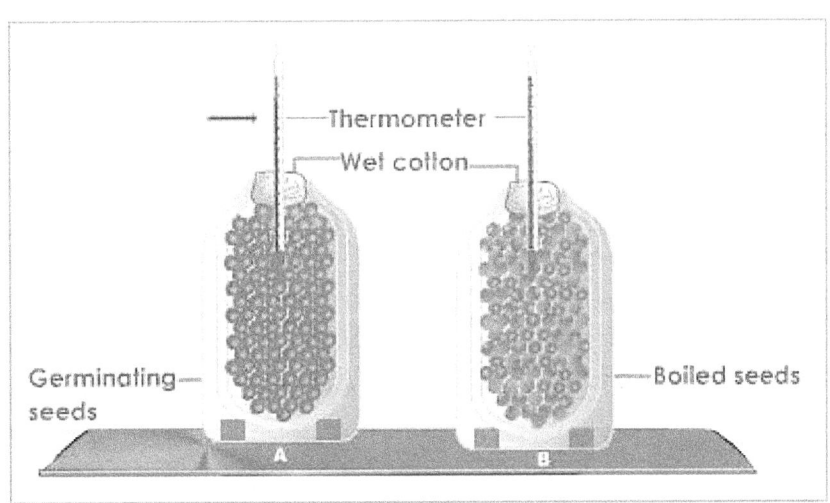

Questions:

1. What is the purpose of respiration?
2. Write the word equation and chemical equation for respiration.
3. Why was one flask filled with boiled germinating peas in the experiment?
4. Explain the temperature changed observed in your results.

Conclusion

Name: _____   Date: _____

Form: _____   Lab Partner: _____

Experiment #: 8

Title: Respiration

Results:

Title: _____

	Temperature before	Temperature after 24hrs
Germinating peas		
Germinating peas (boiled)		

Teacher's Signature: _____   Date: _____

Experiment #: 9

Title: Respiration 2

Aim: To show the structural features of the fish gills

Skill: Drawing

Materials: fish gills, petri dish, pencil, ruler, white paper

Procedure:

1. Place the fish gills in the petri dish.
2. Make a drawing of the fish gills with the longest side vertical on the paper.
3. Label the drawing.
4. Measure the length of the fish gills in the petri dish.
5. Measure the length of your drawing.
6. Calculate the magnification of your drawing.
7. Give your drawing a suitable title.

Name: _____   Date: _____

Form: _____

Experiment #: 9

Title: Respiration

Teacher's Signature: _____   Date: _____

Experiment #: 10

Title: Transpiration

Aim: To plan and design an experiment to investigate the rate at which water is taken up through a leafy shoot under different conditions.

Skill: Plan and Design

Hypothesis:

Materials:

Variable: Manipulating:

              Controlling:

              Responding:

Procedure:

Expected Results:

Treatment of results:

Source of error:

Experiment #: 11

Title: Circulation 1

Aim: To measure the pulse rate of the body.

Skills: Observation, Recording and Reporting; Analysis and Interpretation

Materials: stopwatch

Procedure:

1. Place your middle finger on your wrist and move it about until you can feel your pulse.
2. Count your rest pulse rate for 1 minute.
3. Try doing a little gentle exercise.
4. Measure the pulse rate again for 1 minute.
5. Try exercising a little harder.
6. Measure the pulse rate again for 1 minute.

Questions:

1. Define pulse rate.
2. Explain what happened to the pulse rate after gentle exercise.
3. What effect does harder exercise have on pulse rate? Explain your answer.

Conclusion:

Name: _____   Date: _____

Form: _____   Lab Partner: _____

Experiment #: 11

Title: Circulation

Results:

Title: _____

	Rest	Gentle exercise	Harder exercise
Pulse Rate			

Teacher's Signature: _____   Date: _____

Experiment #: 12

Title: Circulation 2

Aim: To investigate the effect of exercise on heart rate.

Skills: Observation, Recording and Reporting; Analysis and Interpretation

Materials: stopwatch

Procedure:

7. Place your middle finger on your wrist and move it about until you can feel your pulse.
8. Count your rest pulse rate for 30 seconds.
9. Step up onto and down off a step for 5 minutes.
10. Sit down and rest for 1 minute.
11. Measure your pulse rate for 30 seconds and make this value X.
12. Rest for another 30 seconds.
13. Measure your pulse rate for 30 seconds and make this value Y.
14. Rest for another 30 seconds.
15. Measure your pulse rate for 30 seconds and make this value Z.

Questions:

1. How does regular exercise benefit the heart?
2. Why do we breathe more deeply when we begin the exercise?
3. How do the values of X, Y and Z compare to your rest pulse rate?
4. From your results, who is more fit, boys or girls?

Conclusion:

Name: _____   Date: _____

Form: _____   Lab Partner: _____

Experiment #: 12

Title: Circulation

Results:

Title: _____

Student	Resting pulse rate	Pulse rate X	Pulse rate Y	Pulse rate Z	X + Y + Z	Fit / Unfit
Girl						
Boy						

Teacher's Signature: _____   Date: _____

Experiment #: 13

Title: Excretion

Aim: To show the external features of a kidney

Skill: Drawing

Materials: kidney, petri dish, pencil, ruler, white paper

Procedure:

1. Place the kidney in the petri dish.
2. Make a drawing of the kidney.
3. Label the drawing.
4. Measure the length of the kidney in the petri dish.
5. Measure the length of your drawing.
6. Calculate the magnification of your drawing.
7. Give your drawing a suitable title.

Name: _____   Date: _____

Form: _____

Experiment #: 13

Title: Excretion - Kidney

Teacher's Signature: _____   Date: _____

Experiment #: 14

Title: Coordination

Aim: To make a model of the human eye

Skill: Drawing

Materials: embroidery threads, cartridge paper, glue, scissors, buttons, etc.

Procedure:

1. Collect a picture of the human eye.
2. Using embroidery threads and/or cartridge paper, make a model of the human eye.
3. Label and annotate your model.

Name: _____ Date: _____

Form: _____ Lab Partner: _____

Experiment #: 14

Title: Coordination – model of human eye

Teacher's Signature: _____ Date: _____

Experiment #: 15

Title: Sexual Reproduction 1

Aim: To draw the external and internal structure of a kidney bean.

Skill: Drawing

Material: kidney bean, petri dish, scalpel, ruler, pencil, 2 white papers

Procedure:

1. Place the kidney bean in the petri dish.
2. Make a drawing of the external structure of the kidney bean.
3. Label the drawing.
4. Measure the length of the kidney bean in the petri dish.
5. Measure the length of your drawing.
6. Calculate the magnification of your drawing.
7. Give your drawing a suitable title.
8. Use the scalpel to open the kidney bean into half.
9. Make a drawing of one half of the kidney bean to show the internal structures.
10. Repeat steps 3 - 7.

Name: _____ Date: _____

Form: _____

Experiment #: 15

Title: Sexual Reproduction – Red Kidney Bean

Teacher's Signature: _____ Date: _____

Experiment #: 16

Title: Sexual Reproduction 2

Aim: To draw the structure of a half hibiscus flower.

Skill: Drawing

Material: hibiscus flower, petri dish, scalpel, ruler, pencil, 2 white papers

Procedure:

1. Place the hibiscus flower in the petri dish.
2. Use the scalpel to cut the flower into half.
3. Make a drawing of the half hibiscus flower.
4. Label the drawing.
5. Measure the length of the hibiscus flower in the petri dish.
6. Measure the length of your drawing.
7. Calculate the magnification of your drawing.
8. Give your drawing a suitable title.

Name: _____   Date: _____

Form: _____

Experiment #: 16

Title: Sexual Reproduction – Half Flower (Hibiscus)

Teacher's Signature: _____   Date: _____

Experiment #: 17

Title: Asexual Reproduction 1

Aim: To draw the internal structure of an onion.

Skill: Drawing

Material: onion, petri dish, scalpel, ruler, pencil, 2 white papers

Procedure:

1. Use the scalpel to cut the onion length wise.
2. Place one half of the onion in the petri dish, with the inner section facing upwards.
3. Make a drawing of the internal structure of the onion.
4. Label the drawing.
5. Measure the length of the onion in the petri dish.
6. Measure the length of your drawing.
7. Calculate the magnification of your drawing.
8. Give your drawing a suitable title.

Name: _____ Date: _____

Form: _____

Experiment #: 17

Title: Asexual Reproduction - Onion

Teacher's Signature: _____ Date: _____

Experiment #: 18

Title: Asexual Reproduction 2

Aim: To draw the external structure of a ginger.

Skill: Drawing

Material: ginger, petri dish, ruler, pencil, 2 white papers

Procedure:

1. Place the ginger in the petri dish.
2. Make a drawing of the ginger.
3. Label the drawing.
4. Measure the length of the ginger in the petri dish.
5. Measure the length of your drawing.
6. Calculate the magnification of your drawing.
7. Give your drawing a suitable title.

Name: _____   Date: _____

Form: _____

Experiment #: 18

Title: Asexual Reproduction - Ginger

Teacher's Signature: _____   Date: _____

Experiment #: 19

Title: Asexual Reproduction 3

Aim: To draw the external structure of a potato.

Skill: Drawing

Material: potato, petri dish, ruler, pencil, 2 white papers

Procedure:

1. Place the potato in the petri dish.
2. Make a drawing of the potato.
3. Label the drawing.
4. Measure the length of the potato in the petri dish.
5. Measure the length of your drawing.
6. Calculate the magnification of your drawing.
7. Give your drawing a suitable title.

Name: _____   Date: _____

Form: _____

Experiment #: 19

Title: Asexual Reproduction - Potato

Teacher's Signature: _____   Date: _____

Experiment #: 20

Title: Conduction

Aim: To compare the conductivities of different solids

Skills: Manipulation and Measurement; Analysis and Interpretation

Materials: metal rods of different material (copper, iron, steel), glass rod, nail / matchsticks, candle wax, Bunsen burner, tripod stand, stopwatch

Procedure:

1. Rest the rods on a tripod stand and attach a nail or matchstick at one end of each rod using a drop of candle wax.
2. Start a stopwatch and begin to heat the other ends of the rods with a Bunsen burner.
3. Time how long it takes for each nail or matchstick to fall off.
4. Record your results on a bar chart.

Questions:

1. Why are metals better conductors than other solids?
2. From the experiment, which material is the best conductor? Which is the worst?
3. Name one factor that could account for the differences in conductivity?

Conclusion

Name: _____   Date: _____

Form: _____   Lab Partner: _____

Experiment #: 20

Title: Conduction

Results:

Title: _____

Material	Time / s
Glass	
Iron	
Copper	
Steel	
Brass	

Teacher's Signature: _____   Date: _____

Experiment #: 21

Title: Radiation

Aim: To investigate heat of radiation in a shiny and a matt black container

Skills: Manipulation and Measurement; Analysis and Interpretation

Materials: thermometer, hot water, polished and shiny container, matt black container, lid, stopwatch

Procedure:

1. Pour an equal volume of hot water into each can.
2. Close both cans with the lid.
3. Insert the thermometer in the hole on the lid.
4. Stir the water and record the temperature of the water in each can every minute for 15 minutes.
5. On the same axes (graph), plot a graph of temperature against time for each can.

Questions:

1. Which surfaces are the best emitters of radiation? Which surfaces are the best absorbers of radiation?
2. Which can cools down more quickly?
3. Which can is emitting heat energy by radiation more quickly?

Conclusion

Name: _____ Date: _____

Form: _____ Lab Partner: _____

Experiment #: 21

Title: Radiation

Results:

Title: _____

Time / min	Temperature Shiny Can / °C	Temperature Matt Black Can / °C
1		
2		
3		
4		
5		
6		
7		
8		
9		
10		
11		
12		
13		
14		
15		

Teacher's Signature: _____   Date: _____

Experiment #: 22

Title: Terrestrial Environment 1

Aim: To investigate the air content of soil

Skills: Manipulation and Measurement; Analysis and Interpretation

Materials: measuring cylinder, soil, water

Procedure:

1. Measure 50 cm^3 of soil into a 100 cm^3 measuring cylinder.
2. Measure 50 cm^3 of tap water in another measuring cylinder.
3. Add the tap water to the soil in the measuring cylinder and shake the mixture.
4. Measure the combined volume of the soil and water.

Questions:

1. Calculate the volume of air in the original sample of soil.
2. Express the amount of air in the original soil as a percentage.

Conclusion

Name: _____ Date: _____

Form: _____ Lab Partner: _____

Experiment #: 22

Title: Terrestrial Environment

Results:

Title: _____

Amount of Soil	Volume water	Volume of soil and water

Teacher's Signature: _____ Date: _____

Experiment #: 23

Title: Terrestrial Environment 2

Aim: To investigate the water retention in soils

Skills: Manipulation and Measurement; Analysis and Interpretation

Materials: measuring cylinder, different soil types, water

Procedure:

1. Measure 50 cm^3 of each soil type (A, B and C) into a 100 cm^3 measuring cylinder.
2. Measure 50 cm^3 of tap water in another measuring cylinder.
3. Add the tap water to the soils in each measuring cylinder (A, B, C) and shake the mixture.
4. Measure the combined volume of the soil and water.

Questions:

1. Calculate the volume of water retained by each soil type.
2. Express the amount of water retained by each soil type as a percentage.
3. Arrange the soil types A, B and C in decreasing water retention.
4. Identify the soil type of A, B and C.

Conclusion

Name: _____   Date: _____

Form: _____   Lab Partner: _____

Experiment #: 23

Title: Terrestrial Environment

Results:

Title: _____

Soil Type	Amount of soil / cm^3	Volume of water / cm^3	Volume of soil and water / cm^3
A			
B			
C			

Teacher's Signature: _____   Date: _____

Experiment #: 24

Title: Aquatic Environment 1

Aim: To demonstrate the effects of heat on copper sulphate

Skills: Observation, Recording and Reporting

Materials: copper sulphate, boiling tube, Bunsen burner, cobalt chloride paper, spatula

Procedure:

1. Use the spatula and place some copper sulphate into the boiling tube.
2. Heat the copper sulphate gently until no further change occurs.
3. Use the cobalt chloride paper to test the drops of liquid at the top of the boiling tube.
4. Leave the boiling tube to cool.
5. Once the boiling tube is cool add few drops of water to the contents.
6. Note any observations.

Questions:

1. What liquid was given off when copper sulphate was heated?
2. What is the difference between hydrous and anhydrous?
3. Using the terms above, what happened when water was added to the contents of the boiling tube?
4. Suggest why this is described as a reversible reaction.

Conclusion

Name: _____ Date: _____

Form: _____ Lab Partner: _____

Experiment #: 24

Title: Aquatic Environment

Results:

Title: _____

Colour of copper sulphate before heating	Colour of copper sulphate after heating	Colour change of cobalt chloride paper	Colour change of contents after drops of water

Teacher's Signature: _____ Date: _____

Experiment #: 25

Title: Aquatic Environment 2

Aim: To determine the upthrust and density of a stone

Skills: Manipulation and Measurement; Analysis and Interpretation

Materials: stone, balance, cord, two beakers, displacement can, water

Procedure:

1. Find the mass of a dry beaker.
2. Place another beaker under the spout of a displacement can. Fill the can with water until the water runs from the spout.
3. When the water has stopped dripping remove the beaker and replace it with the dried weighed beaker.
4. Tie a string tightly around the stone. Record the weight of the stone in air as displayed on the spring balance.
5. Carefully lower the stone into the displacement can. When it is completely immersed record the new weight displayed on the spring balance.
6. When no more water drips into the beaker, remove the beaker and find the mass of the beaker and water.

Questions:

1. Subtract eh mass of the dry beaker to find the mass of the displaced water.
2. Calculate the weight of the displaced water by multiplying the mass by gravity (10 m/s).
3. Compare the apparent loss in weight of the stone to the weight of water displaced. What conclusion can you make about Archimedes' principle?

Conclusion

Name: _____   Date: _____

Form: _____   Lab Partner: _____

Experiment #: 25

Title: Aquatic Environment

Results:

Title: _____

Mass of beaker / g	Mass of stone /g	Mass of stone immersed / g	Mass of beaker + water / g

Teacher's Signature: _____   Date: _____

Experiment #: 26

Title: Sanitation

Aim: To plan and design an experiment to investigate the conditions in which bread mould grows.

Skill: Plan and Design

Hypothesis:

Materials:

Variable: Manipulating:

　　　　　　Controlling:

　　　　　　Responding:

Procedure:

Expected Results:

Treatment of results:

Source of error:

Experiment #: 27

Title: Safety Hazards

Aim: To demonstrate effect of carbon dioxide on combustion

Skills: Observation, Recording and Reporting

Materials: candle, dilute hydrochloric acid, calcium carbonate, beaker

Procedure:

1. Place the candle in the middle of the beaker and carefully light it.
2. Put some calcium carbonate in the bottom of the beaker.
3. Add a small amount of dilute hydrochloric acid to the beaker and observe what happens.

Questions:

1. What can you say about the density of carbon dioxide compared to air?
2. Does carbon dioxide support combustion? Explain your answer.
3. Name the three components of the fire triangle.
4. How should a fire involving burning oil be extinguished?
5. Which type of fire extinguishers should NOT be used in a fire involving electrical equipment?

Conclusion

Name: _____   Date: _____

Form: _____   Lab Partner: _____

Experiment #: 27

Title: Safety Hazards

Observations:

Title: _____

Test	Observation
Dilute hydrochloric acid added to calcium carbonate	
Candle in the beaker	

Teacher's Signature: _____   Date: _____

Experiment #: 28

Title: Chemical Reactions of Metals

Aim: To investigate the chemical reactions of metals with acid

Skills: Manipulation and Measurement; Observation, Recording and Reporting

Materials: aluminium powder, zinc powder, copper, iron fillings, dilute hydrochloric acid, measuring cylinder, balance, test tube

Procedure:

1. Weigh 0.5g of each aluminium powder, zinc powder, copper and iron fillings.
2. Using a measuring cylinder, pour 5 cm^3 of hydrochloric acid, of the same concentration and temperature, into each of four test tubes.
3. Add a metal powder to each test tube and observe what happens.

Questions:

1. In which test tube were bubbles given off more vigorously?
2. In which test were no bubbles given off?
3. Use your observations to deduce the order of reactivity of these four metals and write them down, starting with the most vigorous.
4. Write the word and chemical equation for the reaction between the four metals and acid.
5. Why is it important to us the same volume of acid, of the same concentration and at the same temperature?

Conclusion

Name: _____   Date: _____

Form: _____   Lab Partner: _____

Experiment #: 28

Title: Chemical Reactions of Metals

Observations:

Title: _____

Metal	Observation with dilute hydrochloric acid
Aluminium	
Zinc	
Copper	
Iron	

Teacher's Signature: _____   Date: _____

Experiment #: 29

Title: Corrosion

Aim: To investigate the conditions needed for rusting

Skills: Observation, Recording and Reporting

Materials: 4 iron nail, 4 test tubes, water, oil, boiled water, anhydrous calcium chloride, sea water or 5% sodium chloride solution

Procedure:

1. Place an iron nail into each of four test tubes labelled A, B, C and D.
2. In test tube A, cover the nail with tap water and leave the tube open to the air.
3. In test tube B, cover the nail with freshly boiled water and cover the water by a layer of oil. Leave the tube open to the air.
4. In test tube C, place small amount of anhydrous calcium chloride and seal the tube with a bung.
5. In test tube D, cover the nail in seawater or 5% sodium chloride solution and leave the tube open to the air.
6. Leave the test tubes for one week and then observe any changes to the nails.

Questions:

1. Distinguish between corrosion and rusting.
2. Write a word equation for rusting.
3. Which tube had the most rust? Why this had the most rust?
4. What is the purpose of the anhydrous calcium carbonate?
5. What is the reason for using oil in test tube B?

Conclusion

Name: _____   Date: _____

Form: _____   Lab Partner: _____

Experiment #: 29

Title: Corrosion

Observations:

Title: _____

Test Tube	Observation after a week
A	
B	
C	
D	

Teacher's Signature: _____   Date: _____

Experiment #: 30

<u>Title:</u> Acids, Bases & Salts

<u>Aim:</u> To investigate the effect of indicators on household items

<u>Skills:</u> Observation, Recording and Reporting

<u>Materials:</u> universal indicator, litmus paper, phenolphthalein, methyl orange, lime juice, salt, toothpaste, milk, vinegar, milk of magnesia, distilled water, cola, oven cleaner, test tubes, droppers

<u>Procedure:</u>

1. Where necessary, make up a solution of the substance by adding a small amount to 1 cm³ of water in a test tube.
2. Add a few drops of universal indicator to each solution.
3. Make a list of the substances and, alongside each one, state whether it is an acid, an alkali or neutral and write its pH value.
4. Repeat steps 2 and 3 using the other indicators. Draw a table of your results.

<u>Questions:</u>

1. Define the terms acid, alkali and salt.
2. What is a pH scale used for?
3. Which indicator gave you more information on the items? Give a reason for your answer.
4. Which item was the most acidic?
5. Which item was the most alkaline?

<u>Conclusion</u>

Name: _____ Date: _____

Form: _____ Lab Partner: _____

Experiment #: 30

Title: Acids, Bases & Salts

Observations:

Title: _____

Household items	Universal indicator (acid/alkali/neutral)	pH value	Litmus paper	Phenolphthalein	Methyl orange

Teacher's Signature: _____ Date: _____

Experiment #: 31

Title: Hardness of water

Aim: To determine the hardness of samples of water

Skills: Observation, Recording and Reporting; Manipulation and Measurement

Materials: rain water, tap water, sea water, soap solution, boiling tube, measuring cylinder, bung

Procedure:

1. Half fill a boiling tube with rain water
2. Add about 3 cm³ of soap solution using a measuring cylinder, cover the end of the tube with a bung and shake the contents for a few seconds.
3. Note your observation and measure the height of the lather if any is present.
4. Repeat steps 1 – 3 using tap water and sea water.

Questions:

1. Define hard water.
2. How can water be softened?
3. Which sample of water was the hardest?
4. Which sample of water was the softest?

Name: _____ Date: _____

Form: _____ Lab Partner: _____

Experiment #: 31

Title: Hardness of water

Observations:

Title: _____

Water sample	Observation	Measurement of lather
Rain water		
Tap water		
Sea water		

Teacher's Signature: _____ Date: _____

Experiment #: 32

<u>Title:</u> Soap and Soapless Detergents

<u>Aim:</u> To make a soap using castor oil

<u>Skills:</u> Observation, Recording and Reporting; Manipulation and Measurement

<u>Materials:</u> sodium hydroxide, castor oil, sodium chloride, cold water, measuring cylinder, beaker, Bunsen burner, tripod stand, glass rod

<u>Procedure:</u>

1. Carefully pour 50 cm^3 of sodium hydroxide solution into a beaker and gently heat it until it nearly boils. Be very careful as this is a strong alkali.
2. Slowly add castor oil and stir continuously.
3. When the oil has disappeared, add one teaspoon of sodium chloride (salt). The soap should precipitate out. If it does not, add more salt.
4. Allow the mixture to cool down and stir it to break up the soap.
5. Filter the mixture through filter paper. Rinse the solid with cold water. This is soap.
6. Add some of the soap to water in a test tube and shake it.

<u>Questions:</u>

1. Distinguish between soap and soapless detergents.
2. Write a word equation for the reaction.
3. Why do some detergents contain enzymes?
4. Why is there less chance of skin irritations with soapy detergents?

<u>Conclusion</u>

Name: _____   Date: _____

Form: _____   Lab Partner: _____

Experiment #: 32

Title: Soap and Soapless Detergents

Observations:

Title: _____

Step	Observation
2	
3	
5	
6	

Teacher's Signature: _____   Date: _____

Experiment #: 33

Title: Current, Voltage & Resistance

Aim: To investigate how different resistances affect current

Skills: Analysis and Interpretation; Manipulation and Measurement

Materials: ammeter, voltmeter, battery supply, variable resistor, connecting wires, test wire, filament bulb

Procedure:

1. Set up the circuit as shown in the diagram below.
2. Adjust the variable resistor to give the lowest current possible though the wire.
3. Record the current on the ammeter and the voltage on the voltmeter.
4. Adjust the variable resistor to give a different value for current and record the current and the voltage.
5. Repeat step 4 until you have five evenly spaced values for current and five corresponding values for voltage.
6. Plot a graph of voltage against current for the wire.
7. Repeat the experiment using a filament bulb in place of the wire.
8. Plot the results on the same graph from above.

Diagram:

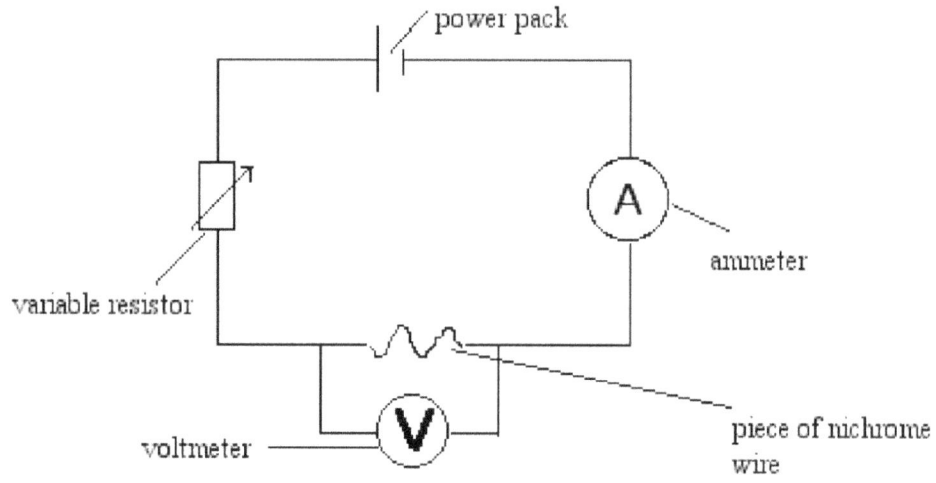

Questions:

1. Define the terms current, voltage and resistance.
2. State the formula that relates current, voltage and resistance.
3. What conclusion can you make from your graphs?
4. Select one set of reading from your result and calculate the resistance.

Conclusion

Name: _____   Date: _____

Form: _____   Lab Partner: _____

Experiment #: 33

Title: Current, voltage & resistance

Results:

Title: _____

Test wire		Filament bulb	
Current / I (A)	Voltage / V (V)	Current / I (A)	Voltage / V (V)

Teacher's Signature: _____   Date: _____

www.ingramcontent.com/pod-product-compliance
Lightning Source LLC
Chambersburg PA
CBHW080537190526
45169CB00007B/2529